LONDON BABY STATIONS

DEDICATION: TO MAMA

LONDON BABY Stations
ISBN: 9781739211585
Published by LANDE Jewels
LANDE Jewels turns mundane pastimes into exciting and informative adventures.
www.landejewels.com

 A CIP catalogue record for this book is available from the British Library.

THIS READING JOURNEY BELONGS TO

Contents

Stations

Tickets
Northern line

Prologue

Uniqueness of the London stations,
Its pioneers, historic fame,
Was subject of the conversations
That mum insisted we maintain.

From high tech modern architecture
To listed features of brickwork,
The London infrastructure network
Offers unrivalled culture shock.

Important urban education
That every Londoner must have
Is how to navigate locations,
Where to get in, how to get out,
What to do during signal failures,
Which route to take during planned works,
When to respect rush-hour queuing,
How to sneak in the closing doors.

As one embarks on thrilling journey
Of love-hate blunders of commute,
Appreciation starts its showing—
It isn't perfect, it is good!

TRANSPORT TRUST
METROPOLITAN
RAILWAY
The world's first underground railway
opened from Paddington to Farringdon
via Baker Street Station on
10th January 1863
For further information visit
www.transportheritage.com
TRANSPORT HERITAGE SITE

Beginnings

The web of different colour lines
With knots of interchange
Is labelled with familiar words
That come to fascinate.

From East to West and North to South,
Deep down and over ground,
The transport network duly serves
Commuters across town.

When first ambitious plan was put
For Parliament's review,
Rejection followed, yet two groups
Merged and had scheme approved.

Charles Pearson, father of the tube,
Died months before the launch
Of Metropolitan Railway's
Four miles of tunnel jaunt.

From terminus at Paddington,
Covering seven stops,
Steam train arrived at Farringdon
To shareholders' awe.

Bakerloo Line

Humour enticing portmanteau*,
Dropped in a press report,
Was picked up, humoured ever more,
And commonly adopted.

The route had a dramatic start—
After initial flops,
The mining mogul pledged the funds,
Yet took his life in courts.

Charles Yerkes resumed the masterplan
With add-ons West and East,
Relieved congested Met at last
In swarming Baker Street.

Thanks to receiving little love
Over some previous decades,
The trains, well past their shelf life,
Retain the charm and quaintness.

Art-Nouveau decorative tiles,
Red brick with stone detailing,
'No smoking' roundels, signal hubs
Make vintage stations—epic.

*Portmanteau refers to two words combined to form a new one—Baker Street and Waterloo combination formed Bakerloo.

Central Line

The Central London Railway
From Bank to Shepherds Bush
Opened at turn of century,
Nicknamed two-penny tube.

In days before the travel card,
It meant instant success—
The economic and fast ride
Left travellers impressed.

It's built directly under roads
To overcome concerns
Of toppling grandiose St Paul's
And listed streetscape core.

It was extended West and East;
Disused Ongar in Essex
Remains the 'zero-kilometre'
For network distance measure.

The hot and squeaky, poorly lit,
Underground's longest line,
Awaits built-in CCTV
To curtail lawless acts.

Circle Line

Line was a venture between rival railways
Who fought and struggled to cooperate,
Yet forced to carry on by Act of Parliament—
Met—clockwise, and District—anticlockwise.

Not separate until the WW2 postbellum,
Storied as Inner Circle, Circle Train,
Thought of as route with first class carriage travel
Made frugal due to armed conflict outbreak.

Logistic headache, it was sometimes threatened
With closure to make way for other lines.
Instead, new terminus gave it a breather,
Replacing circle with a spiral run.

It scores two unique ways, it's loved by tourists,
It's famous in pop culture for pub crawl,
For legendary corpse no one has spotted,
For signal failures at Edgware Road.

District Line

The second oldest line with the most stations
Was opened on a frosty Christmas Eve.
Linking, for openers, five destinations,
It now has sixty stops for public use.

It was the first to welcome female drivers,
It closed on Sundays for church-hours break,
It barred fishmongers from moving their wares
On air-conditioned, electrified trains.

A very complex web working six branches
Crosses railways, tube lines and Overground.
An intense schedule for the present signals
Means trains come into view vaguely on time.

Green line uniquely crosses Thames on Bridges,
Serves thirteen boroughs in six fare zones,
Tops Overground bizarrely at Whitechapel,
Bamboozles travellers awaiting at Earl's Court.

A nonexistent house in Leinster Gardens
Covers the tracks with a Victorian facade.
Nearby Porchester Terrace panorama
Offers compelling peak at truth behind.

Docklands Light Rail (DLR)

A child of noughties, hero of the Docklands,
That had revived the derelict terrain
With only fifteen stations when it opened,
Eleven cars, two paths and solid aims.

Grown to six routes, it reached forty five stations.
Run by the private sector as franchise,
It doesn't share track with other railways;
No ticket barriers require reader touch.

Second light rail in Britain has no driver,
The job is filled by kids of any age,
Thrilled by a trip akin to fairground riding,
Deep down at Bank, high up at Tower Gate.

Emerging from the darkest City tunnels,
It meets East London expanse of low rise,
Meanders through the valley of skyscrapers,
Skirts stadiums and the Olympic Park.

Cars shake and wobble as wheels hunt for balance.
They cross the Thames and meet Regent's Canal,
Offer a glimpse of some abandoned platforms,
Pass celebrated prime meridian.

Elizabeth Line

When £15 billion Crossrail project
Eventually opened its gates,
The extra money and the time spent
Seemed rather easy to forget.

The purple train, named after monarch,
Running on current and new tracks,
Enthused a nation of anoraks
With extra long walkthrough layout.

Its awkwardness is in its essence—
Short metro service plus commute
Means usual comfort essentials
Are not included, far removed.

Sobering hardness of the seating
Is softened by a soundless glide.
The science-fiction stations' outfits
In uniformity stand out.

The largest UK excavation
Struck resting souls of Black Death pit,
Objects from past civilisations,
And creatures long ago extinct.

Hammersmith & City Line

Born to feed into Metropolitan railway,
With an extension into Hammersmith suburb,
The pink line, frankly separate in name,
Shares track and all twenty five stops.

The row between Great Western and the Met,
That nearly put an end to H&C,
Achieved an amicable end result—
Money was made and line was launched in peace.

The route relieves the bottlenecks off neighbours
In a reliable and comfortable stock.
The seat moquette reflects where it is rolling;
Timetable is sound and signals work.

1. Stratford
4. Stratford
12:24:08

Jubilee Line

The planning and construction phase of Fleet Line,
Referencing a long lost river Fleet,
Was finalised to stop at Charing Cross end
Renamed after Queen's Silver Jubilee.

There were, back then, protesters with objections—
"Would Jubileeve it?" they complained of costs.
Yet, mostly with old stations' infrastructure,
It came as new and functionally strong.

The Northern part carved out of Bakerloo line
Helped balance flow during the jamming peaks.
Further extension eastward to the Docklands
Delivered volumes to financial district.

Other projections never saw fruition—
Neither Ludgate Hill nor the Fenchurch Street
Were given go ahead to grey line stretching;
At Charing Cross, the platforms went extinct.

A satisfying melody of train huff
Is getting louder as the stock adds age
Thanks to thyristors regulating current
As it's picked up from electrified rail.

Metropolitan Line

Efficient cure to capital congestion,
"Insult to common sense" The Times announced,
Immediately advantageous venture,
Brought fascination with underground lines.

The Metropolitan Railway organisation
Aimed at long distance routes in Metroland,
Targeted wealthier gardened specimens,
Offered a Pullman beside regular ride.

Worlds oldest line after consolidation
Retained some attitude, autonomy, design—
The transverse seating, finish, views of landscapes,
The space for luggage and umbrella rack.

Slowly, the Met was brought in line with others,
Its journey—shortened, tall speed limit—cut.
High back sprung seats were lost to frugal ambiance—
The bench-like hardness is to be endured.

The banter by the Buckinghamshire locals
Points to prestige nostalgia of the route
And argues it's not tube but the mainline,
Akin to Thameslink or Chiltern commute.

Northern line
Edgware branch
Platform
1

Northern Line

Steady supply of horror stories
Kept fresh the folklore of the line—
With out-of-date signal equipment,
It suffered regular breakdowns.

Mind-boggling junctions and extensions
Maintain uniqueness of its paths—
Two different routes through Central London,
Two branches North, two branches South.

It was suggested line was split up
In cost effective route upgrade,
Necessitating full remaking
Of Camden Town logistics maze.

The conservators barred the scheming
And after decade of no-news,
Battersea Power stop extension
Revived the long-forgotten woos.

'Misery Line' nickname, black colour,
'One-under' record, Moorgate crash,
Kennington Loop ghost, cast-off tunnels—
The Northern Line runs unabashed.

Overground

An orbital commuter network,
That's also known as ginger line,
Forms part of National Rail service,
Serving around M25.

Built almost wholly on the old track
With 19th century layout,
It employed Marc Brunel's Thames Tunnel,
Revived after the fires and floods.

Unused during the Thatcher years,
The ghost railway made sound comeback,
Connecting isolated areas
With opportunities 'off-plan'.

The expectation of the money
Caused massive demolition spree—
Template developments for hipsters,
Dwarfed by tall structures, made the scene.

Revamp has tamed what was once seedy,
Disparate lines greatly improved,
Gentrification and Olympics
Pushed into spotlight current routes.

PICCADILLY RLY

Piccadilly Line

The longest line following Central,
Cockforsters-Uxbridge (or Heathrow),
Adopted by US financier
With not a single line, but four.

Meandering in road plan fashion,
Avoiding easements as it curves,
It covers over fifty stations
Plus, seven more, counting disused.

Some in the zone of Central London
Were built with shelters and blast walls—
Down Street safeguarded Winston Churchill
Throughout continuous raids of war.

The Aldwych-Holborn branch projection
Encouraged West End show pastime.
In wartime, it stored museum relics;
In peacetime, though, it didn't thrive.

Gone are the days of theatre shuttle,
Hail to extensions and upgrades.
Stretching conveniently to airport,
It's full of people with suitcase.

Tramlink

Gentrification of South London
Has been a rumour for some time.
An introduction of tram service
Was tantamount to speed things up.

Connecting Wimbledon and Croydon,
Lime coloured light rail found new use
For long abandoned railway track lines
That were transformed for gentler cruise.

Fast, well designed mass transit system
Is powered by overhead wires.
It's more efficient than the buses
For its capacity and size.

It's built to navigate tight corners,
Accelerate, brake in a trice.
A tragic accident, however
Imposed strict rules and sign guidelines.

A scenic glider cuts through parkland,
Backyards and gardens, quiet roads,
Meets busy high street, gas works towers,
Rubs shoulders with main network stock.

← Lift
to all areas
← Victoria Coach Station
← Way out
National Rail

Victoria Line

An automated, fast and frequent service,
With trains at less than two minutes apart,
Fit above Piccadilly, below Metropolitan,
From start to end entirely under ground.

The largest challenge was the Oxford Circus
Where steel umbrella kept traffic intact
While works below were rapidly evolving—
Concourses, escalator shafts.

King's Cross kept running as the works proceeded,
Enlarging station, altering its shape;
The clay-rich soil gulped tunnelling equipment
And integrated reinforced concrete.

A curiosity between King's Cross and Euston
Allows to travel northbound in both ways—
Head on following strictly one direction
And end up at exact beginning place.

Uniquely pictured platforms feature motifs
(In seating recess) by renowned artistes.
The stations' elevation slows approaches
And aids acceleration in a blink.

↑ Waterloo & City line

Waterloo & City Line

The shortest line with just two stops—
The Waterloo and Bank,
Offers unique and bumpy ride
As it dips sharply down.

The Londoners call it 'The Drain'
For dampness, rusty smell—
Millions of litres every day
Leak through the tunnel shell.

Removal of the ampersand
From its name acronym,
And rumoured place for gents above,
Bolster the reasons still.

Purchased for nominal £1
In privatisation wave,
Line joined the underground, at last,
Having been British Rail.

Healthy demand keeps route alive
Into and from Square Mile.
Flushed of its workers on Friday night,
On weekend it welcomes stars.

Kingsway Tramway

The Kingsway Tramway subway tunnel
That city merrily embraced
Was an ambitious stretch of tram track
In flourish until bus progressed.

Labour intensive cut-and-cover
Produced a cutting edge result.
Further advance to lull and glamour
Failed to compete with metro cult.

One of its kind service was short-lived.
With less than fifty years in,
It was abandoned, the tracks—lifted,
Trams—scrapped, replaced by diesel din.

Part of the subway had a face lift,
Appearing as Strand Underpass,
Reducing traffic of the junction,
It serves as corridor for cars.

The listed heritage, however,
Is shown in transport museum tours.
The closed off sections offer storage
For spare signs and litter bins.

WESTMINSTER BRIDGE HOUSE

London Necropolis Railway

In early nineteenth century,
Numbers of London dwellers—
Doubled, yet space for graveyards
Was tightly regulated.

A booming corpse collection
Contaminated water,
Prompted the epidemics—
Cholera, smallpox, typhoid.

It was, therefore, decided
To make accommodation
For all future dead bodies
In out of town location.

A funeral train service
To Brookwood burial site
Carried coffins and carriages
In dedicated cars.

Depending on religion,
The class, material wealth,
The dead would be assorted
To their respective place.

London Necropolis Railway *contd.*

The grounds were so enormous,
The scheme underachieved—
It saw much fewer bodies
That settled in its pits.

When WW2 had started,
Bombs fell on Waterloo
And damaged the Necropolis
Putting it out of use.

The cost of reconstruction
Failed to uphold the cause,
The journey to eternity
Now had to be by road.

Moreover, British Railways,
In March of '88
Formally ceased to carry
The bodies of the dead.

The cemetery, however,
To this day is going strong.
Served by South Western Railway,
It welcomes the footfall.

Mail Rail

The overcrowding of Victorian London
Led to proposals of underground post—
The system of conveying mail in tunnels
With air pressure as propelling force.

The WW1 put breaks on its completion,
Resuming in late twenties with a launch—
The route between Paddington and Whitechapel,
Eight stations running nearly round the clock.

World's first electric railway without drivers,
In time, became fully computerised.
Its halcyon ended with switch to text message,
Contrastive costs meant it could not survive.

Abandoned, it was partly resurrected
As a museum, venue and film set.
Eery, yet fascinating public section
Sheds light on flood doors, cables, forlorn stock.

Accessible through Mount Pleasant Mail Centre,
The place where a notorious English jail
Neighboured a dump heap so unpleasant
That irony translated into name.

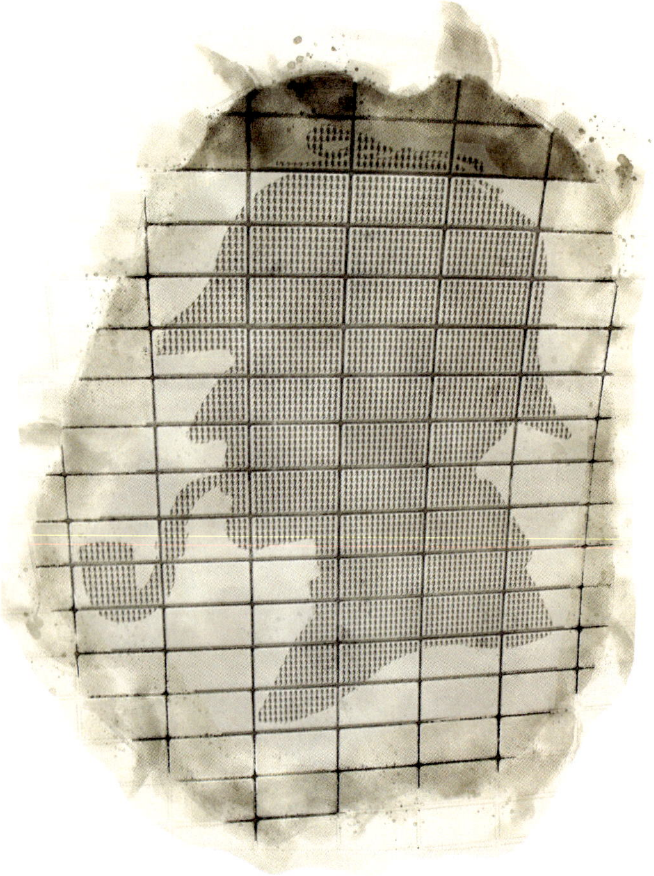

Baker Street Station

One of the first underground mentions,
Connected to five subway lines,
With most platforms of any station,
It's dotted with old plans and plaques.

Primaeval cherished approaches
Are shorter than more recent league.
The drivers, therefore, warn voyagers
That front and back doors wouldn't clear.

The alcoves in cavernous structure
Let steam from engine to escape
In days when Metropolitan Railway
Was burning coal to haul the trains.

The closed-off corridors and passes
Alongside the disused lift shafts
Reveal array of social spaces
And even rifle range for staff.

The silhouette of sharp detective
And illustrated famous scenes
Emblazon Jubilee Line sections,
Attracting daily shutter clicks.

PRINCES
STREET
UNDERGROUND

Bank Station

With its foundations in mass grave
And Central Bank on top,
Colossal tunnel corridors
Imbibe the platform curves.

The warning message 'MIND THE GAP',
Coined in the 1960's,
Particularly holds at Bank
Where gap is quite malicious.

Commuter-crowded concrete maze,
Length of King William Street,
Has ventilation shafts extend
To statues overhead.

United duo Monument-Bank
With nine lifts, sixteen exits,
Boasts travolators, ten platforms,
Thirty-plus escalators.

The Black Nun, ghost who groans and wails,
Blitz bomb, churchyard, plague pit,
Help to relate the tomblike air
Of long transfigured crypt.

Blackfriars Station

Originally called St Paul's,
The mainline railway station,
Took Blackfriars as official name
To thwart wrong tube connection.

Supporting the commuter trains
Which were revived as Thameslink,
The termini over decades
Absorbed substantial goods freight.

The platforms straddle River Thames
Across Blackfriars Bridge,
Displaying copious rebuild
Next to red iron piers.

Hydraulic buffers were removed,
Distinctive roof—replaced,
The customary office block
Struck as austere place.

The destination names stone plaques
Were deep cleaned and installed
Atop the wall of the concourse
Of modern tube-rail store.

UNDERGROUND
CAMDEN TOWN STATION

Camden Town Station

V-shaped design of surface building
Tells little of the complex web
Of hypogeal trailblazing
That copes with daily flow and ebb.

Northbound—directly above Southbound,
Four tunnels cached beneath High Street,
Linked Edgware and Chalk Farm branches—
Shows arduousness of precise fit.

Century-old, it needs attention
To curb capacity constraints,
As growing man manifestation
Blocks exits and the interchange.

Attempts to remedy the problem
Required large scale clear out.
The City Hall prompt intervention
Insisted on less drastic plan.

Critical infrastructure project
Lacks funding for the go-ahead.
Despite appealing new proposals,
The public is waiting for consent.

Canary Wharf Station

Canary Wharf had made the headlines
As futuristic and well-morphed
Phenomenon with gullwing ceiling,
By oval columns held aloft.

With background of expanding site,
The station box was neatly settled
Into an ex water-filled basin,
When dried and emptied—just right size.

The talk of water reinstatement
Proved to be technically unwise,
Except for glass canopied entries,
It's largely hidden under park.

Example of piecemeal construction,
Canary Wharf remains disjoined—
The DLR and Crossrail sections
Are ways off Jubilee platforms.

Despite the navigation hurdles,
It scores well in design and art—
Enormous roundels, lush roof garden,
Culture events and screens onsite.

Cannon Street Station

A product of Victorian railway rivalry
That warranted construction of new bridge,
It was setup atop medieval steel yard
And served as Charing Cross efficient link.

Impressive City skyline contribution,
A glaring target during WW2,
Accents two noble weathervane topped features
That contrast distant fairness of St Paul's.

The post war real estate boom raised new questions
About bomb-damaged, funding-hungry scene,
Opening up commercial exploitation
Of air-rights as sound revenue stream.

Further corporate use of station landscape
Included a squash club and office blocks
With flat roof garden to maintain perspective
From Tower Bridge to the Cathedral dome.

Restricted hours for regular commuters
Were ended with addition of Thameslink.
The underground District and Circle service
Have allied both timetables in a tick.

Charing Cross Station

Charing Cross Station

Arrivals to the London Bridge
Were headed to West End;
Convenient station was approved
And prospered straight away.

It was main continental hub
For traffic of boat-trains,
Important military point
For wartime personnel.

Proposal to have station moved
To cheap land further South
Wasn't supported and had lapsed;
Soon, WW2 began.

Since bombing during Hitler's War,
Site underwent development
With offices 'Embankment Place'
And footbridge over river Thames.

Its front looks nothing like its back—
The French Renaissance style
Contrasts with air-space robot-like,
Bulky Post-Modern sight.

CHARING CROSS STATION *contd.*

A tragedy is in its name—
The Queen Eleanor,
En route to join Edward I,
Died suddenly; King mourned.

He ordered a memorial cross
Erected for his wife
In dozen pass-through villages,
Charing—being the last.

Original structure was destroyed
During the Civil War.
The courtyard replica is one
Of that atop Whitehall.

A 'Central London' for the cabs
Refers to five mile radius
Around the Charing Cross (Strand side),
Examined during Knowledge Test.

Westbound
platform 3

Earl's Court Station

The home of first underground escalator,
Suspicious stairwell travelling up and down;
Whose rescue came from one-legged Bumper Harris,
Who rode all day to help clear the doubts.

Applauded for its architectural significance,
Barry's glass roof spanning all four platforms
Lets rare sunshine on District and Piccadilly lines
Recurrently on split level concourse.

The current station is a new location,
The first one perished in sweltering flames.
It keeps intact important interchanges,
Takes credit for District Railway success.

Historic destination board, light boxes
Were lovingly restored by TFL*,
Linked to the latest signalling equipment
During the ever-quiet owl shifts.

An overwhelming station for first timers,
It teaches to locate the clue displays
And follow clear old school arrows
To get to the right train without delays.

*TFL-Transport For London

UNDERGROUND

Embankment Station

Location with confusing name relations,
Originally built as Charing Cross,
Was called four decades later the Embankment;
Title continued changing back and forth.

Thanks to the mayhem of competing railways,
More stations were approved into the mix—
Trafalgar Square, Strand, Charing Cross Embankment
Have merged and split into the current fix.

Removed from Strand by a steep slope, it borders
Lush, decorated gardens on one side
And modernist construction on the other,
With bridge above it stretching to South Bank.

Inside, there hides a mystery substation,
Huge, empty space behind blast door—
Pages Walk tunnel, thought as haunted
Due to cold spots and sense of being watched.

Signature vintage 'Mind the Gap' recording
Was reinstated after the upgrade
When station members honoured a love story—
A wish to hear late husband intonate.

Euston
Euston
Euston Station

Euston Station

An early main-line termini
Created the upheaval
When tracks cut through the fields and streets
Frightening semi-villagers.

Named after Duke's home—Euston Hall,
(He owned much local land),
Euston became the place first shaft
Was sunk into the ground.

Moving though muck beneath New Road,
Building the first tube trench,
The 'cut and cover' tunnelling
Soon let in swift steam trains.

The architectural grandeur—
Hardwick Hall, marble floor,
Doric archway—seemed purpose-flawed,
Expensive, massive, tall.

Historic structures were destroyed
By British Rail establishment
In favour of current eye-sore
With basic comforts absences—
Station approach through taxi rank,
Dingy bus stops and plaza,
Harrowing platforms, dire ramps,
Intense traveller chaos.

Euston Station *contd.*

Remaining features of the past,
The ornate gates and lodges,
Father of railways set in bronze—
Help tolerate the glumness.

Hiding below the station grounds—
Layers of derelict tunnels,
Treasure trove of the artefacts,
Sealed off with cabled passages.

Rapidly forming High Speed Rail
With Euston 3 as terminus,
Promises praiseworthy design,
Future proofing modernity.

Large demolition is alive,
Clawing in nearby complexes.
Thousands of crusty, crumbled tombs
Head to Brockwood to rest in peace.

Fenchurch Street Station

The first railway termini
Within the City walls
Connected ancient Minories
With West India Docks.

The trains were dragged by station staff
Uphill from Blackwall base,
And let to follow downward path
With a momentum haze.

Abandoning the cable-haul,
Station adopted steam.
With a few platforms, it absorbed
Frenzied commuter stream.

Victorian frontage was maintained
Despite major rebuild;
A high rise block in place of roof
Crowned newly wired strip.

Considered to form part of Fleet,
Now called Jubilee Line,
It's not yet linked to underground,
Yet, it is close enough.

Hampstead Station

The deepest station due to its advantage
Of sitting at the top of a steep hill,
At nearly sixty meters below daylight
Showcases tiles with former name—Heath Street.

Twenty five storey high dramatic staircase
Spirals into tormenting soot and dust,
Intoxicating fitness-seeking climbers,
Keen to take up the challenge if they must.

The seasoned travellers proceed to elevators
Through corridors lit by fluorescent lights,
And rise into a charming booking section
With welcome bench and history insights.

UNDERGROUND
HOLBORN STATION

Holborn Station

Arrival of the Belle Epoque
Brought with it groundwork plans—
The end of rookeries and slums
Was firmly in the cards.

Clearance required transport links
For workers to commute—
New tramway underneath Kingsway
And new underground route.

To bring to life an interchange
Of Piccadilly-Central,
Hand excavation went ahead
Next to the busy train tracks.

The planning rules put breaks on tiles—
Holborn has stone facade.
Enamelled metal murals art
Enriches platform side.

Station revamp is coming soon,
Subject to fitting purse.
Deserted platforms five and six
May, at last, go for good.

Holland Park Station

A keen observer would discover
The concrete render over tiles
In unambiguous biscuit colour,
Once, terracotta shade half-fired.

Inside, the herringbone floor pattern,
The platform clock and old type signs,
Cast iron columns holding ceiling,
Plaster where platforms were enlarged.

A generous Jacobean structure
With spacious grounds—the Holland House,
Has largely vanished during wartime
While station in its name survived.

King's Cross Station

The folklore tale reveals a story
Of bloody battle between Romans
And Celtic British tribe Iceni,
Led by Boudicca who fell there—
Around the Platform eight and nine,
Where travellers report the sights
Of fearsome spirit, warrior ghost,
At home within London's King's Cross.

The most connected London station
Derives its name from the location
Where short lived statue of King George
Marked intersection of three roads.

Built as a terminus for Northern Railway,
It handled people, goods, livestock and freight,
Facilitated passenger locomotive getaway
On Flying Scotsman and a record-breaking Mallard.

It pioneered the smoke-free public transport travel
After the match set devastating wooden escalator fire.
It was the first to have a Platform nought
And feature Europe's largest single span concourse.

Liverpool Street Station

Replacement to the Bishopsgate,
Great Eastern Rail depot,
The 'black cathedral' of steam trains
Disturbed old sculls and bones—
Bedlam asylum burial ground
Atop a Great Plaque slew,
Rested in peace until the chiefs
Let loose the digging tools.

Construction cleared notorious slums,
Which stretched far into Shoreditch.
Suburban traffic in and out
Helped Corporation flourish.

Hit by WW1 air raid,
It soon was back in service.
It welcomed Jewish refugees
As part of kindertransport.

The noughties' facial revamp
With conservation triumph,
Allies tube, airport express
And complex Crossrail tunnels.

London Bridge Station

The only station that contains
London in its full name
Is also oldest termini
Still running to this day.

Modest ambition to construct
A London-Greenwich line,
Railway on brick-built viaduct,
Made a historic mark.

The structure—sixty million bricks,
With tree-lined promenade,
Delivered novel spectacle
Of nightly lit lampshades.

London Bridge station with no roof
And three exacting tracks
Became the target of rife bout
For access into town.

Five kilometres of arch-run,
The longest in UK,
Was widened to the North and South—
To foil future dissent.

London Bridge Station *contd.*

Competing companies paid tolls
Until rates got too high;
Resulting separate approach
Brought chaos to both sides.

War damage added to the maze,
Such that final development,
Provoked by Jubilee expanse,
Resolved signage and faring.

A goldmine of the artefacts,
Dawned from underground digging,
Saw bronze age loom weight, pots and tiles,
And crumbling Roman pits.

The arches, previously, homes,
Stables, workshops and pubs,
Became abyssal catacombs
Of passenger footpaths.

Victorian-era church organ
Resides on the concourse.
Named 'Henry', it was saved and homed
For its cascading tone.

Marylebone Station

The last Victorian main line into London,
Built as a termini by Great Central Railway,
It faced the fiercest contest for the custom,
Struggled to yield commercial success.

Sir Edward Watson—mogul of train travel,
Dreamed of Manchester-Paris splendid route;
The area was cleared and redeveloped
With full regard towards the neighbour's turf.

Demure, somewhat provincial construction,
Barren from Marylebone Road by grand hotel,
Remembers porters who outnumbered passengers,
While fortunes were arriving through goods freight.

It prospered after British Rail formation—
Prestigious locomotives made a sight,
The Wembley Stadium extension drew in large crowds,
And further goodwill came from underground.

The terminal faced closure in the 80's.
It, luckily, escaped with big upgrade.
Improved commuting, hybrid power sections
Made choosing it—an affable crusade.

OXFORD
CIRCUS
HOUSE
CIRCUS

Oxford Circus Station

No matter reconstructions,
Extensions, stairs, shafts,
The station is, undoubtedly,
Is never big enough.

Historically, two stations
Were beautifully designed,
Except for askew tunnelling
With subway paths off-plan.

Arrangement of two railways,
Logistics of mains, sewers,
Restrictive clauses of the land
Meant limited manoeuvring.

Addition of Victoria Line
Was so large in its scale
That steel umbrella heavyweight
Framed traffic for five years.

The formidable labyrinth
Scoffs thick dynamic crowds
That flow in exits, entrances
Through X-crossing above.

1

Paddington Station

The best preserved of the Victorian terminals,
One out of seven first stops on the tube,
Paddington—Bishops Road initially,
Is famous for its triple-span glass roof.

Born from Brunel-Wyatt collaboration—
Great Exhibition hands-on expertise,
Church-like layout enclosed in giant train shed,
Had Great Western Royal as street facade.

Expensive land meant only bare necessities.
Logistics, though, were excellent and sound,
With link to the canal for shipments,
And end of line for Shillibeer's bus.

An early 20th century expansion
Added the perishable market traffic,
And tonnes of daily issued printed press,
Destined for delegation to Mail Rail.

Adorned with sculptures and memento plaques,
A painted bench and Paddington Bear mount,
It prompts first timers and the regulars alike
To echo the astounding WOW!

PUBLIC
UNDERGROUND
SUBWAY

Piccadilly Circus Station

The station, fully underground,
An engineering marvel,
Underwent plenary revamp
To cope with greater numbers—
Anteros statue was removed,
The fountain stored in depot,
While eighteen foot wide shaft was sunk
With junction undisturbed.

Business as usual was maintained
Despite extensive works,
Gestating seamless upgrade launch
After four-year course.

Cream marble cladding, bronze, woodwork,
Art-deco lights, passimeter,
Neat escalators and phone booths,
Spoke loudly of bohemia;
Revealing opulence and pomp,
Befitting flagship station,
Which, to this day retains its charm
As tourist destination.

A world map feature and artwork
In honour of Frank Pick,
Prompts pondering and taking note
Of covered district.

Sloane Square Station

Connecting Chelsea with the wider network,
Heaving with people during Flower Show,
The station opens to sought-after shopping
Of tinselled Sloane Street and swinging King's Road.

Victorian engineers solved the challenge
Of culverting the Westbourne from the tracks
With gravity storm combo system sewer
That's seen as a green conduit above.

Remarkably, the darling was survivor
Of heavy bombing during the Great War.
While canopy and platforms turned to rubble,
The river-sewer bided undisturbed.

Hole in the Wall pub at the platform level
With Kim—resident buffet cat,
Helped quench the thirst of regular commuters
Well into noughties when it closed at last.

Postwar design is functional, lacklustre,
Olive green tiles—replaced with white and blue,
The garden trellis pattern lost to charmless,
Characterless, utilitarian, astute.

St Pancras International Station

Romantic and spectacular masterpiece—
The train shed and hotel of largest scale,
Was born of architect and engineer
Harmoniously mixing their trade.

The booking hall is set with detailed drainpipes,
The undercroft—converted to concourse,
Arcaded galleries resemble travel walkways
Complete with pianos and passenger lounge.

The Midland Grand Hotel—Renaissance,
Features Medieval stairwell, iron beams.
It narrowly escaped the demolition
When Eurostar construction plans came in.

The station's permanence is owed to a campaigner
Sir John Betjeman who insisted to preserve
Domestic rail hub with ornate interiors—
He strove not to demolish but conserve.

Named after child-saint who died by beheading
For wanting to retain his Christian faith,
The terminus hosts hives of pollinators
That make delicious honey on its slates.

London Victoria Station

Victoria Station

Located next to the Buck House,
Convenient termini for the West End—
Victoria, a monument to rivalry
Of pioneer commercial railways.

Built as two stations alongside each other—
Edwardian Baroque on 'Brighton side',
Next to contrasting Portland stone on 'Chatham's',
Physically disconnected and distinctly run.

Eventually, the railway grouping,
Brought down the walls between the two.
Newly exposed eye-pleasing arches
Added to glamour of detours.

Rapid success of air-rail service
Ended the 'Golden Arrow' days;
While Southern Belle train since the 30's
Maintains its mile per minute pace.

The office blocks and retail outlets,
Remodelled bus stops in forecourt,
Accompany tired-looking station,
Letting down heritage on show.

WATERLOO STATION

Waterloo Station

Formerly, a Nine Elms extension
For Central London passengers and freight,
The most confusing of Victorian termini,
Evolved into the best laid in UK.

The viaduct connection scarred the parishes
Of Lambeth, Vauxhall, Battersea,
Dissecting landmarks and displacing houses,
While keeping major monuments pristine.

Completed station underwent transitions—
Following bombings during WW2,
It lost Necropolis, gained Jubilee extension,
Converted Eurostar, added Crossrail.

Time traveller remembers hordes of porters
Meeting Southampton boat-train gadabouts,
Arrival of Afro-Carribean migrants,
Departures and returns of Bournemouth Belle.

Hydraulic buffers iconize the platforms,
Waterloo clock remains a meeting place,
Britannia is presiding over entrance
While tens of millions pass through every year.

Westminster Station

A modern, unassuming narrow entrance
Opens to steel and concrete warehouse space.
The genius of civil engineering—
Construction on industrial scale, no less.

Great care was taken during excavation
Of deep line subterranean railway;
Precise as not to undermine foundations
Of Abbey, Church and Parliament terrains.

As trains got longer, tunnels were extended,
Followed by monstrous works on Jubilee—
Unfathomable dig in Central London,
A gaping void for station carcass fit.

Proximity of tunnels to clock tower
Presented real risk of structural cracks.
The mounting of the steel tubes, grout affixing
Were used to militate against collapse.

Praised for the architecture and aesthetics,
The platform safety, ease of rush-hour flow,
It serves a busy tourist destination,
State epicentre and religion's lore.

Enjoyed the book?

Please, dear reader, stay in touch,
Leave a review or two, share much.
And on that note—goodbye, adieu,
I will be glad to hear from you.

www.landejewels.com/contact

OTHER BOOKS IN THE SERIES

LONDON BABY CHRISTMAS - ISBN: 9798665093406

LONDON BABY RIDES - ISBN: 9798666032565

LONDON BABY SEASIDE - ISBN: 9798665082868

LONDON BABY SIGHTS - ISBN: 9781739211547

LONDON BABY BELLS - ISBN: 9781739211516

LONDON BABY MARKETS - ISBN: 9781739211523

LONDON BABY BIRTHDAY - ISBN: 9781739211530

LONDON BABY WILDLIFE - ISBN: 9781739211554

LONDON BABY STREETS - ISBN: 9781739211578

LONDON BABY STATIONS - ISBN: 9781739211585

LONDON BABY BRIDGES - ISBN: 9781739211592

ALSO AVAILABLE AS EBOOKS

* 9 7 8 1 7 3 9 2 1 1 5 8 5 *

BV - #0062 - 150524 - C51 - 203/127/7 - PB - 9781739211585 - Gloss Lamination